MARION CONSTRUCTION MACHINERY 1884 THROUGH 1975 PHOTO ARCHIVE

Including Shovels, Draglines, Backhoes, Clamshells, Cranes, Log Loaders and Railroad Ditchers

Edited by the Historical Construction Equipment Association

Iconografix
Photo Archive Series

Iconografix
PO Box 446
Hudson, Wisconsin 54016 USA

© 2002 Historical Construction Equipment Association

All rights reserved. No part of this work may be reproduced or used in any form by any means... graphic, electronic, or mechanical, including photocopying, recording, taping, or any other information storage and retrieval system... without written permission of the publisher.

The information in this book is true and complete to the best of our knowledge. All recommendations are made without any guarantee on the part of the author or Publisher, who also disclaim any liability incurred in connection with the use of this data or specific details.

We acknowledge that certain words, such as model names and designations, mentioned herein are the property of the trademark holder. We use them for purposes of identification only. This is not an official publication.

Iconografix books are offered at a discount when sold in quantity for promotional use. Businesses or organizations seeking details should write to the Marketing Department, Iconografix, at the above address.

Library of Congress Card Number: 2001135742

ISBN 1-58388-060-7

02 03 04 05 06 07 08 5 4 3 2 1

Printed in China

Cover and book design by Shawn Glidden

Copyediting by Suzie Helberg

COVER PHOTO: See page 61.

Book Proposals

Iconografix is a publishing company specializing in books for transportation enthusiasts. We publish in a number of different areas, including Automobiles, Auto Racing, Buses, Construction Equipment, Emergency Equipment, Farming Equipment, Railroads & Trucks. The Iconografix imprint is constantly growing and expanding into new subject areas.

Authors, editors, and knowledgeable enthusiasts in the field of transportation history are invited to contact the Editorial Department at Iconografix, Inc., PO Box 446, Hudson, WI 54016.

About the Historical Construction Equipment Association

The Historical Construction Equipment Association (HCEA), a registered, non-profit organization founded in 1986, is dedicated to preserving the history of construction, surface mining, and dredging equipment. This includes corporate archives, histories of product development and use, descriptive literature, memorabilia, and, of course, the machines themselves. Members hailing from all corners of the world are bound together by their interest in historic construction equipment and a concern for the preservation of these machines and their histories.

One of the most important activities of the HCEA is the publication of a quarterly magazine, *Equipment Echoes*. Featuring a color cover, this magazine contains historical, educational and technical articles, numerous beautiful photographs, and reports on historical equipment news and events.

The HCEA also sponsors annual conventions at various locations in North America. These conventions include displays and demonstrations of historical equipment; historical presentations and movies, toy and memorabilia displays and dealers; and plenty of friendship with people who share a common love for old machinery.

The HCEA has established the world's first public archive and museum, The National Construction Equipment Museum, dedicated exclusively to the history of the construction, dredging, and surface mining industries. Located near Bowling Green, Ohio, the facility is open to the public by arrangement. Nearly 2,000 companies are represented in the archive's holdings, which include extensive collections from Marion Power Shovel, Euclid, Galion, Terex, Bucyrus-Erie, Caterpillar, and others. The museum houses over forty machines, some of which are fully restored and operable, and also includes a complete restoration shop.

If you want to become a member of this growing organization or need more information about the Historical Construction Equipment Association, visit our website at www.hcea.net, call (419) 352-5616, or email at hcea@wcnet.org.

Acknowledgments

The Historical Construction Equipment Association wishes to acknowledge the following members for their help in bringing this book to fruition: HCEA Managing Director, Don Frantz for researching and writing the photo captions; HCEA National Director, Keith Haddock for writing the introductory history and editing the captions for historical and technical accuracy; HCEA Archivist, Tom Berry for scanning all of the photographs and editing the text; and HCEA National Director, Harry E. Young for helping with the photo selection.

Marion Power Shovel Company Corporate History

by Keith Haddock

The famous Marion Power Shovel Company had its origins in the 1880s when shovel operator Henry M. Barnhart became so frustrated by breakages and delays due to design flaws in the primitive machines of the day that he conceived a new type of shovel to overcome these deficiencies. In need of financing and a place to build his machine, he turned to industrialist Edward Huber of Marion, Ohio. Huber was impressed with Barnhart's shovel idea and immediately secured a joint patent with him in 1883. Known as "Barnhart's Steam Shovel and Wrecking Car," the first shovel was built in the Huber shops and sold to the Jackson & Mackinaw Railroad Company.

Mr. Barnhart operated and tested the shovel the first season. He and Huber were so sure of its potential that in August 1884 they founded the Marion Steam Shovel Company, along with another friend and industrialist, George W. King. By the time three shovels had been erected in the Huber shops, the young company was able to move into its own new factory in Marion, Ohio, a site which would remain the company's home for the rest of its life.

New products soon emerged from the fledgling company. The first was the ballast unloader in 1885. This simple device resembled a plowshare, and was drawn along a train of flatcars to discharge ballast where needed. The following year Barnhart designed and built a log loader in conjunction with the Goodyear Lumber Company of Buffalo, New York. The loader was mounted on a railroad flatcar, and featured a full-circle swing using a live roller circle. It was claimed to be the first full-circle machine of its type in North America. Barnhart's Railroad Ditcher soon followed. This was built on similar principles to the log loader but was equipped with a shovel attachment. The machine could clean and widen ditches on both sides of the railroad track, and its full-circle capability was a major advantage for loading adjacent railroad cars. Its ability to travel on rails mounted on standard railroad flatcars further increased its effectiveness. The Railroad Ditcher product remained in the Marion line until 1919, and its design was used as a basis for Marion's later fully-revolving shovels and draglines.

Railroad construction formed the larger part of all excavation work in the nineteenth and early twentieth centuries, and the products manufactured by Marion during this period were geared to this market. Marion's steam shovels progressed from Barnhart's early design to a full range of shovels by 1888. The company's steam shovel production received a boost during the construction of the Panama Canal when no less than 24 Marion shovels were shipped to that prestigious project. Like the original Barnhart, all steam shovels up to this time were of the railroad type; i.e. they ran on standard-gauge railroad track, and their booms and dipper attachments were capable of swinging only about 180 degrees.

Marion's first fully-revolving shovel, the Model 30, appeared in 1908. Carrying a 7/8-cubic-yard dipper, it was an immediate success, and the larger 1 1/4-cubic-yard Model 35 was introduced that same year. Revolving shovels gradually overtook production of the railroad shovels, and with the introduction of the heavy-duty quarry and mine shovels in the 1920s, eclipsed the railroad shovels altogether. Marion shipped its last railroad shovel, a Model 61, in 1931.

Crawler tracks first appeared on a Marion shovel, a Model 28, in 1916. For a number of years, Marion offered either crawler or traction wheel mountings for its small revolving shovels, but contractors soon realized the advantages of crawlers, even with their higher prices, and wheel mountings became obsolete.

The Marion company grew into one of the foremost manufacturers of excavating machines and, in competition with its archrival Bucyrus, made similar products such as railroad shovels, dredges, cranes, walking draglines and drills. It changed its name to the Marion Power Shovel Company in 1946, when steam shovels had not been produced for well over a decade. Marion's vast excavator range extended from the smallest 1/2-yard shovel to the largest ever put to work, the famous Captain Model 6360 with an almost unbelievable dipper capacity of 180 cubic yards!

Throughout the 1950s and early 1960s, Marion expanded its construction-sized excavator business with some important acquisitions. It purchased the Osgood Company and its subsidiaries, the General Excavator Company and Commercial Steel Casting Company, both of Marion, Ohio, in 1954, and acquired the Quick Way Truck Shovel Company of Denver, Colorado, in 1961. These acquisitions enabled Marion to offer a broader range of cranes and excavators.

Later in the 1960s, Marion gradually pulled away from the small machine market, preferring to concentrate on walking draglines, blast hole drills, and large two-crawler excavators to serve the booming mining industry.

In 1997, Marion Power Shovel Company was purchased by archrival Bucyrus International, Inc., formerly Bucyrus-Erie Company. The coming together of these two giants was a significant event in the earthmoving industry, and abruptly ended an intense competitive rivalry spanning 113 years. The plant at Marion, Ohio, closed, but certain machines from the former Marion line have been updated and are still available today as Bucyrus machines.

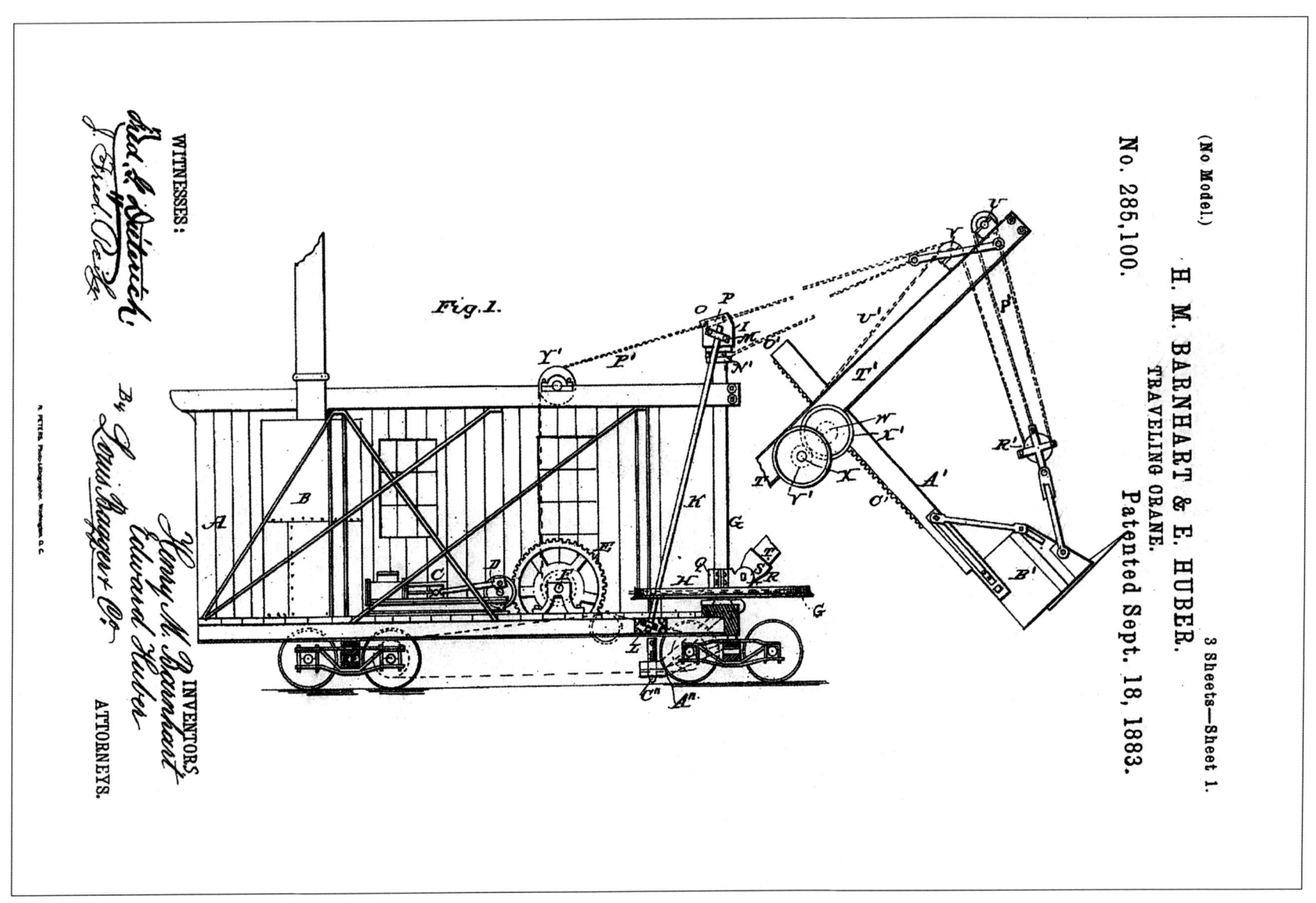

Patent drawing for the first Marion steam shovel designed by Henry M. Barnhart and manufactured by Edward Huber. The Marion Steam Shovel Company was incorporated in August 1884.

"Steam shovel No. 2," now designated a Style "A." The dipper could be removed allowing the machine to be used as a railroad wrecker to put railroad cars and locomotives back on the track. Four hundred sixteen models were built through 1906, making it easily the most popular of the early shovels.

This 1 1/4-cubic-yard Style "A" was sold to the E.H. France Company of Bloomville, Ohio, in 1886. HCEA member Don Frantz's grandfather was an employee and may have operated this machine in 1913. Note that the propelling drive is now located externally.

A venerable "Barnhart Special" railroad shovel with its horizontal boiler protruding from the rear. Note the water tank attached to the rear of the shovel.

The H.S. Kerbaugh Company owned this Style "C" seen here in Philadelphia, Pennsylvania. Thirty-four of these 3/4-cubic-yard machines were built from 1888 through 1907. Note the large external flywheel.

The 2 1/2-cubic-yard Style "AA" was offered from 1891 through 1897, with 74 being built. The craneman can be clearly seen standing on the swing circle. His job was to operate the crowd function and pull the rope on the dipper trip.

This 1896 photograph of a Barnhart Style "AA" was taken at a phosphate mine in Bartow, Florida. Working with Marion engineers, this company has built one of the first backhoes, fittingly called a "back-acting" shovel!

Only one example of the Model "H" was built. This 3-cubic-yard machine was sold to the Western Dredging & Improvement Company in 1896, and worked at Chappel Station, Illinois.

The Model "G" replaced the "AA" in 1897. This improved 2 1/2-cubic-yard shovel was offered through 1902 with 124 examples built. In this 1901 photograph note the beautiful cast builder's plate at the rear of this machine owned by the Patton & Gibson Company.

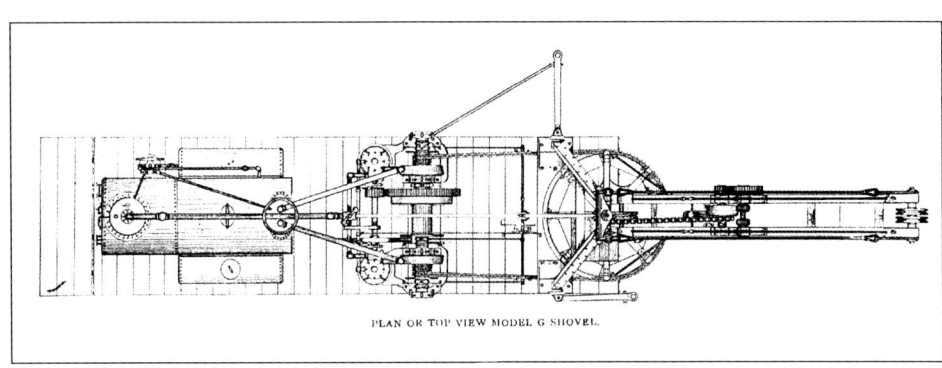
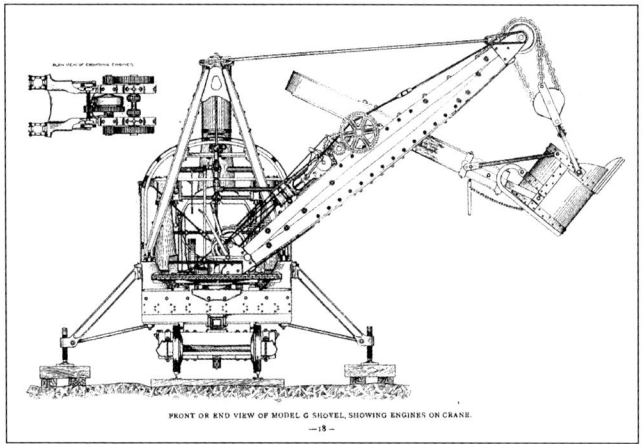
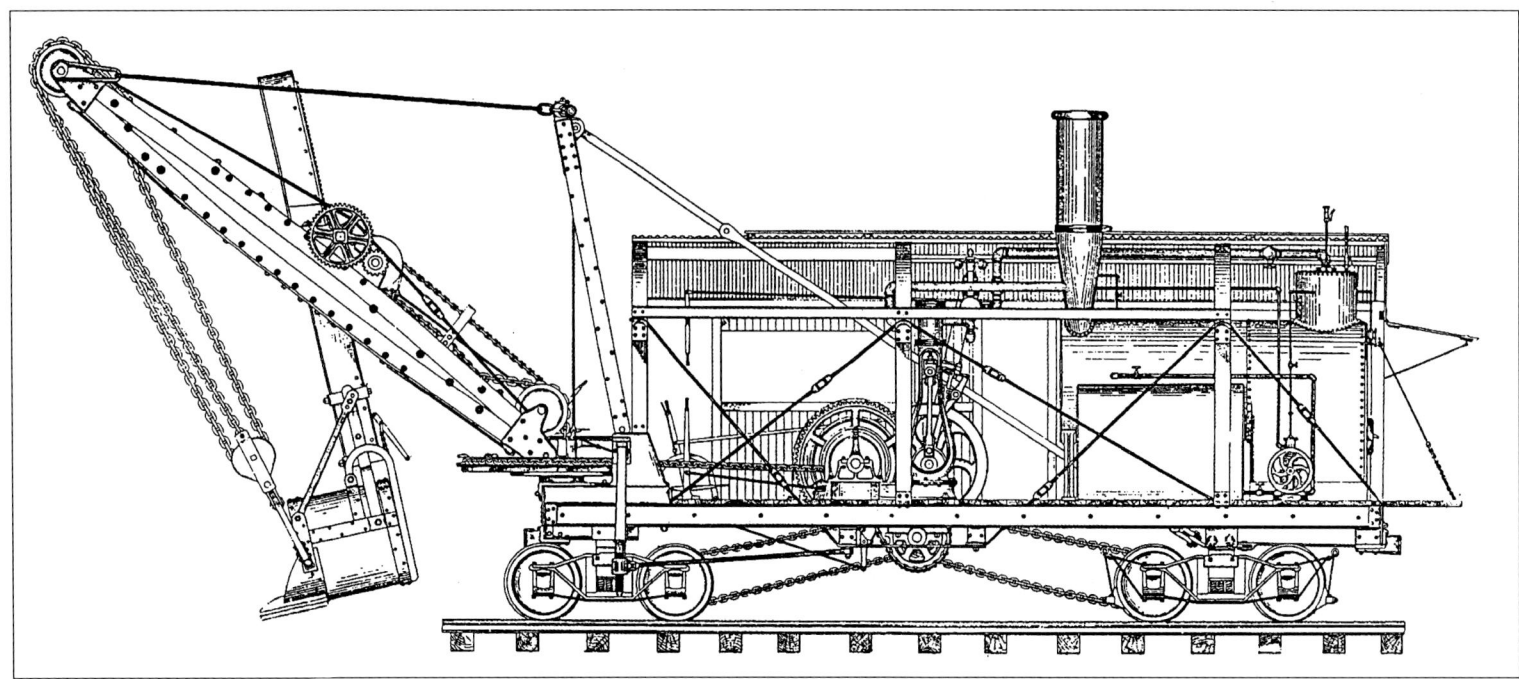

Taken from an 1897 catalog, these line drawings of a Model "G" provide a good view of the construction of a typical Marion shovel of this era.

Judging by the size of the rocks, dynamite must have been at a premium! This Model "K" boasted a 3-cubic-yard dipper capacity but its lifting capacity was not listed. The Model "K" was the first Marion model to incorporate separate engines for each major function. Nineteen of these machines were built between 1898 and 1900.

The Marion Steam Shovel Company abandoned letter designations in favor of model numbers in 1900. These models were designated roughly by gross weight in tons. Forty-six of the 3 1/2-cubic-yard Model 80s were built by 1905.

This 1 1/4-cubic-yard Model 20, one of 228 built between 1901 and 1912, is seen in the factory yard awaiting delivery to the Conner Lumber & Sand Company of Laona, Wisconsin. Dipper swing on most shovels was over 120 degrees.

This Model 20 looks to be ready to "move up." While the fireman relaxes after getting up a full head of steam, the craneman can be clearly seen with crowd throttle in his left hand and dump rope in his right.

Replacing the successful Type "G" in 1902, the mid-sized Model 60 was offered through 1912 with 613 machines built, making it the most popular railroad shovel ever offered by Marion. The engineer (or shovel runner), craneman, and fireman pose proudly.

Perhaps one of the most famous Marion railroad shovels was the Model 91, built from 1902 to 1912. Of the 131 built, 16 of these 5-cubic-yard machines were supplied to the Isthmian Canal Commission (ICC). This May 2, 1912, photograph shows a proud shovel and pit crew after a record-setting shift excavating on the Panama Canal.

This 1909 photograph shows a special electric Model 91 being built for the Casparis Stone Company of Columbus, Ohio, for a job in Sanborn, New York.

The 2-cubic-yard Model 50 was introduced in 1904 with 54 units being built by 1912. This photograph was taken in 1906 in Knoxville, Tennessee. A common construction feature of most railroad shovels was for the rear wall of the "house" to drop and serve as a platform to hold coal or wood for fuel.

The 4-cubic-yard Marion Model 75 was built from 1907 through 1912, with 46 examples constructed. An easy way to distinguish between Bucyrus and Marion railroad shovels is by their boom supports. Bucyrus used massive straps while Marion used the forged "hog rods" as seen here.

Many of Marion's smaller and mid-sized railroad shovels were fitted with traction wheels. This eliminated the need for rails, ties, and the large pit crew required for moving-up. This 1 1/4-cubic-yard Model 40, offered from 1908 to 1912, is loading a woefully inadequate horse-drawn dump wagon for contractors Conover and Taber.

This Model 40 has its work cut out as it digs through un-blasted shale at East Radford, Virginia. The crew looks perfectly resigned to its fate!

In 1909, Marion introduced its 5-cubic-yard Model 92, essentially replacing the older Model 91. Offered until 1930, it was one of the last of the railroad shovel models.

Beginning in 1923, Marion offered crawler trucks to replace railroad trucks on railroad shovels. As can be seen in this photograph of a Model 92, a pair of crawlers replaced the rear truck under the house, and crawlers were placed in front at the ends of each outrigger.

Another Model 92 has been equipped with crawlers. This photograph offers a clear view of the front crawler attached to the outrigger. As a result of the addition of crawlers, the life span of railroad shovels was greatly lengthened, with many working into the 1950s.

Also introduced in 1909, the 6-cubic-yard Model 100 was the largest Marion railroad shovel produced. By the time it was discontinued in 1926, 39 units had been built. The open door offers a view of one of the massive 13x16 hoisting engines.

Filling a niche for small railroad shovels, the 3/4-cubic-yard Marion Model 25 was built from 1910 through 1921. This machine was built in 1910 for the Rinehart & Dennis Company of Valhalla, New York.

This diminutive Model 25 was specially constructed for tunnel work. Note the large reservoir for compressed air. Three Model 36 full-revolving traction machines are parked in the background.

Between 1911 and 1929, Marion constructed 65 of these 4-cubic-yard Model 76s. This 1916 photograph of a Winston-Dear Company shovel shows a deep cut being made near Keewatin, Minnesota.

This Model 41 is seen loading dump cars on a railroad grade separation job. This craneman, who has a clear vantage point on the side of the boom, has to work very closely with the engineer who controlled the hoisting and swinging functions of the shovel. Crowd engine exhaust is emitting from the end of the boom. Seventy-five of these 1 1/2-cubic-yard machines were built between 1912 and 1923.

The 2 1/2-cubic-yard Model 61 replaced the Model 60 in 1912, with 131 units built. When discontinued in 1931, it was the last railroad shovel built by the Marion Steam Shovel Company.

The 3 1/4-cubic-yard Model 70 saw 74 units built between 1912 and 1930. The massive hoisting chain seen here is only mid-sized for railroad shovels—1 3/8 inches! All railroad shovel booms were made of oak with steel cladding. The Model 70 shown here is working at New Haven, Connecticut.

The 2-cubic-yard Marion Model 51 replaced the Model 50 in 1913. This represented the last new railroad shovel model offered by Marion, and 22 units were built by 1928. This electric machine was built for the Empire Limestone Company of Buffalo, New York.

Although Marion had full-revolving log loaders in the 1880s, it was not until 1908 that it introduced the 3/4-cubic-yard Model 30, Marion's first full-revolving excavator. Produced until 1912, the 54 models built could be mounted on traction wheels or rail mounted as pictured. This machine has an exceptionally long dipper for use in trenching work.

This 1912 photograph is of the same Model 30 machine (No. 2477) as in the previous photograph but now equipped with traction wheels. Trenching with shovels always looks precarious.

39

The 1 1/4-cubic-yard Model 35 was also produced from 1908 to 1912 with 53 units built. This 1911 photograph taken in Uruguay shows a rail-mounted shovel owned by the Pan American Transcontinental Railway Company. Notice the rail clamps and outriggers.

This Model 35 is loading side dump cars for the City of Toronto in August 1911. Looks like two buckets to a car!

The 5/8-cubic-yard Model 28 was a very successful early full-revolving excavator with 402 units built by 1919. Traction wheels offered mobility unattainable with railroad shovels, but staying on top of the ground was always a problem. Planks and blocking were the order of the day.

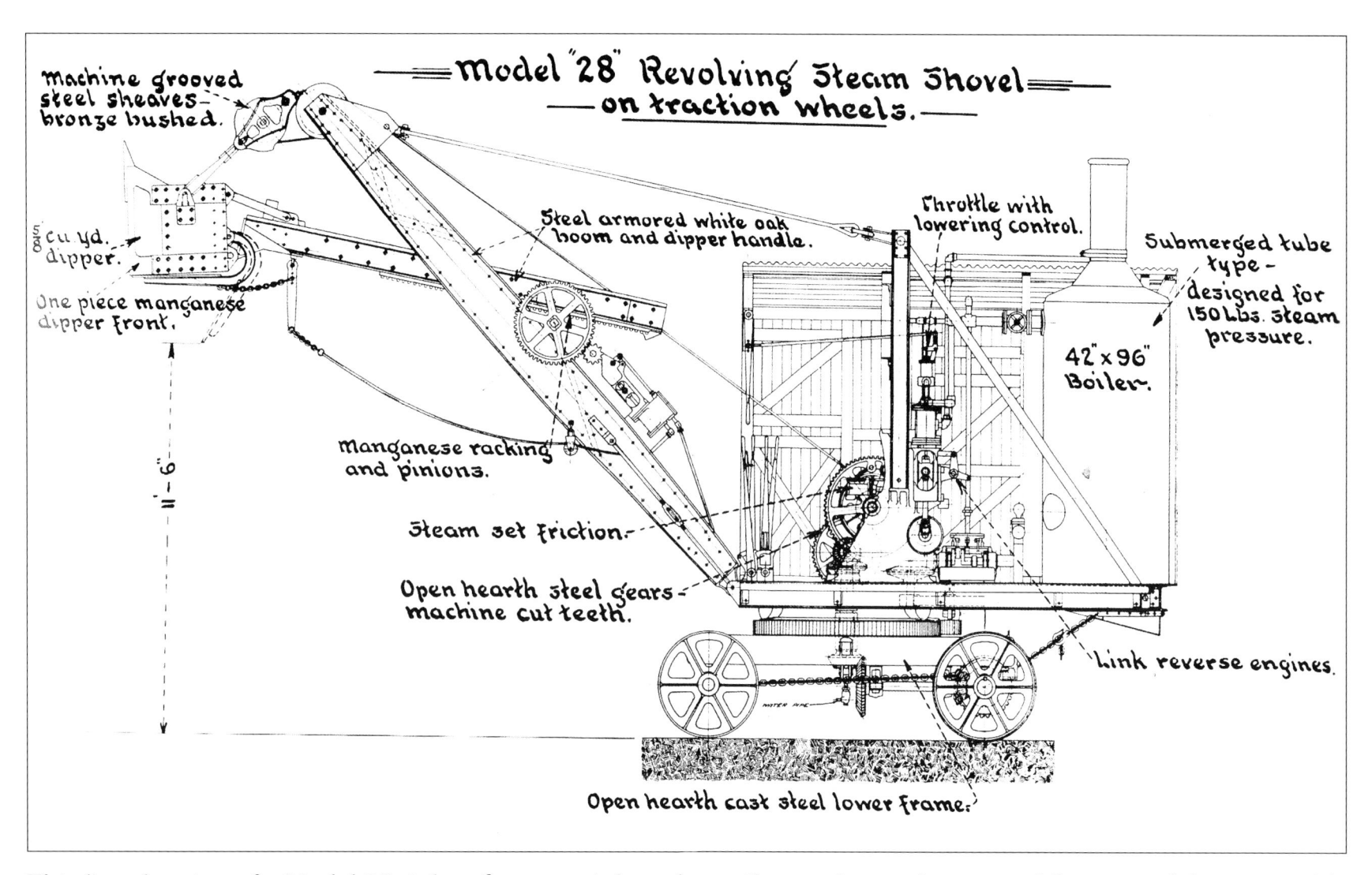

This line drawing of a Model 28, taken from a catalog, shows the machinery layout and features of this venerable machine.

About 1916, Marion added crawlers to a Model 28, making it the first Marion machine so equipped. This Gallatin, Ohio, county dragline is doing a fine job cleaning the ditch but the road is going to be difficult to traverse!

Because a shovel always has to move forward when digging, when a contractor used a shovel for ditching the machine had to straddle the cut. This always led to interesting adaptations. This Model 28 at Youngstown, Ohio, looks like it is sitting pretty. The long dipper is for digging well below grade and casting spoil far to the side.

The Model 250, introduced in 1911, was Marion's first stripping shovel. With a 3 1/2-cubic-yard capacity, this giant was also used by construction contractors for large earthmoving contracts. Note the churn drills on the bank drilling blast holes to loosen the soil.

Introduced in 1912, the 1-cubic-yard Model 31 replaced the Model 30 in the Marion line. Two hundred seventy-nine of these models were built by 1923 when production ceased. This factory photograph shows a traction crane equipped with a Hayward orange peel bucket. Eventually, the Model 31 became Marion's first diesel-powered excavator.

This Marion Steam Shovel Company factory photograph shows a brand new Model 31 with its stepped roof to accommodate the tall boiler.

In 1912, Marion began introducing a line of large "scraper bucket" or dragline excavators. The first of these was the 3 1/2-cubic-yard Model 260.

Also introduced in 1912, the 1 1/2-cubic-yard Model 36 replaced the Model 35. The Western Wheeled Scraper Company of Aurora, Illinois, was the world's foremost producer of railroad dump cars of all sizes.

This Model 36 traction shovel is working its way downgrade. This looks like a typical mining community with its look-a-like company houses. How do you spell "Pittsburgh?"

This 1919 photograph of an early crawler-mounted Model 36 was used in a Marion ad and carried the caption, "Take the short cut to the next job!" The dog seems considerably less interested than the rest of the onlookers.

This factory photograph shows a fully exposed Model 36 electric. Extra weight will be added to the rear-end casting to make up for the lack of weight resulting from the absence of a boiler, water tank, and coal bunk.

The Model 220 dragline was produced from 1913 to 1915, with only five being built. Skids, rollers, and winches were used for propulsion. The machine shown here is working in Spokane, Washington.

The Model 251 had a 3 1/2-cubic-yard capacity and weighed in around 150 tons. Replacing the Model 250 in 1913, just 10 were produced up to 1915.

Only three of Marion's Model 261s were produced. This machine is mounted as a clamshell dredge and is swinging a 3 1/2-cubic-yard bucket.

Marion only built one 5-cubic-yard Model 281 and sold it as a stripping shovel. As evidenced here, at some point later in life the dipper was removed and the machine was equipped as a clamshell dredge. Notice the rail mounting on the barge.

In 1916 Marion introduced its Model 221, 2-cubic-yard dragline. Only two of these models were produced by 1917.

The 3/4-cubic-yard Model 21 was easily the most popular early excavator produced by Marion. Between 1919 and 1926 approximately 810 models were built. This 1922 photograph shows Marion's new logo and its unique four-crawler option on a machine owned by the city of Spokane, Washington.

The Model 21 could be powered by steam, gasoline, electric, or gasoline-electric. This four-crawler machine was owned by the Byrne Brothers of Chicago, Illinois.

Basement excavation in Akron, Ohio, by the Akron Storage & Contracting Company. This Model 21 is equipped with the early version of full-length crawlers.

This 1925 photograph shows a Model 21 dragline at work in Pueblo, Colorado. The Model 21 was the first Marion machine to use single steel casting for its base, thus ensuring positive alignment of shafts and bushings.

This gasoline-electric Model 21 is seen working in August 1926 near Rochester, New York. Marion employed a small four-cylinder gas engine directly connected to a 25 kilowatt DC generator. It is also equipped with Marion's revised full-length crawler.

Marion's 1 1/2-cubic-yard Model 32 was built from 1922 through 1933. The steam shovel's endless appetite for water is being served by the siphon hose running from the wagon (lower right) and up through the machine's center of rotation. This job is at Ashville, North Carolina.

This 1926 factory view of a Marion Model 32 undercarriage offers a sense of the amount of massive castings used in early excavators.

This early gasoline-electric Model 32 is seen working in 1922. Travel gears are all open and exposed to the elements and job hazards.

This 1 3/4-cubic-yard Model 37 shovel is fitted with Marion's unique four-crawler mounting. Introduced in 1922, 330 machines were built by 1929.

A Marion Model 37 dragline digs in Monroe, Louisiana, in 1924. The house has been extended on both sides for shade and storage.

This 1926 photograph shows an electric Model 37 clamming material for the Montgomery Gravel Company, Montgomery, Alabama.

This Marion Model 37 electric shovel is seen digging in Birmingham, Alabama, in 1924. The operator is standing at the controls of the Ward-Leonard DC control system used by Marion Steam Shovel Company starting in 1919.

Only one 3 1/2-cubic-yard Model 262 dragline was built in 1922. It is seen here digging for the D.B. Hearin & Son Company. This photograph provides a clear view of the railroad tracks and rails used for mobility.

In 1925, Marion built its only Model 222, a 3-cubic-yard dragline. All of the steam boilers used in Marion's early stripping shovels and draglines pushed the envelope for hand firing.

By 1929, Marion's only Model 222 had been converted to electric power and had also received a new, all-metal house.

Commencing in 1926 with the introduction of the Type 7, Marion dropped the designation "Model." Two hundred forty-seven of these 1-cubic-yard machines were built through 1929.

This gasoline-electric powered Type 7 dragline is doing what draglines do best, digging ditches! Type 7s were also powered by steam, gasoline, electric, and diesel.

The Type 7 was one of Marion's first backhoe-equipped machines. This gasoline-electric machine is seen digging a ditch in July 1927.

The Driscoll Trucking Company has its hands full trying to move this Type 7. No less than three trucks are required to keep this load moving!

A well-worn Marion Type 7 is seen dumping a full clam bucket. Note the gravity-operated tagline mechanism on the boom used to counterweight the bucket's swing.

In 1927, Marion introduced the 1-cubic-yard Type 440, replacement for the Type 7. These machines were offered with gasoline or diesel power. One can see the traveling counterweight on the base of the boom acting as a tagline for the clam bucket.

The Marion Type 460, built between 1927 and 1934, was only offered with electric power. Only 26 units of this model were built by Marion.

This 1 1/4-cubic-yard Type 450 is loading an early cab-over Autocar truck. The Type 450 was built from 1928 to 1944.

The Type 450 was one of the first Marion cranes to be equipped with a box-section lattice boom. This operator is just about "two-blocked" with his clam bucket against the tip sheaves of the crane.

This 1929 photograph offers the viewer a good look at the front end of an early Marion Type 450 backhoe. The dual diameter drum on the base of the boom was a means for multiplying the line pull on the drag cable.

This Type 450 diesel-electric machine is seen working for the Rhoads Contracting Company, Ashland, Pennsylvania. The view through the sliding door shows the resistor bank needed for the DC power.

One hundred forty Marion Type 480 machines were built between 1928 and 1944. These 2-cubic-yard machines were offered as steam, diesel, or electric powered.

This 1929 factory photograph shows a Marion Type 480 being steamed-up for run-in before the wooden house has been added. This remarkable view shows all!

Another view of a brand new Type 480 being put through its paces at the Marion factory before delivery. Note the well-built wooden house.

Great Lakes Construction Company of Cleveland, Ohio, is using their Marion Type 480 crane to drive wood piling. Notice the wood piling stacked on the bank and the many small stakes marking the piling's locations. The extra tall boiler was required to supply steam for the steam hammer.

The 1-cubic-yard Marion Model 340 pictured above was introduced in 1931 and replaced the Model 440. Marion finally got it right when they introduced the Model 340's replacement, the Model 342, in 1938, and built 159 units by 1947.

Marion's attempt at a half-yard machine resulted in the Model 120 in 1931. Its unique features included hoisting drums at the rear of the house and a large, horizontal swing gear extending over the full width of the crawlers. Note the crawler brake "dogs" which drop down into the crawler shoes.

This side view of a Model 120 shows the crawler sideframes which were very similar to those used by several other manufacturers, including Hanson Clutch & Machinery Company, Ohio Locomotive Crane Company, and the Universal Power Shovel Company (Unit). The 120's radical design caused it to be withdrawn from production the year after it was introduced.

In 1935, Marion introduced a new 3/4-cubic-yard excavator, the Type 331. It was produced through 1947 with 364 machines built.

In 1936, Marion introduced its 3-cubic-yard Type 39-A. The Type 39-A was quickly replaced in 1937 by the nearly identical 40-A. Both machines were offered only as draglines.

Marion replaced its 1933 line of clutch-type excavators in 1938. The first of the new line machines, the 1 3/4-cubic-yard Type 372, actually came out in 1937. The new line featured clutches controlled by vacuum valves rather than the former manual type. This 372 is having its lifting capacity seriously tested!

The new Marion Type 362 replaced the Type 361 in 1938. The series II Type 362, introduced in 1942, represented the same machine but with a few more improvements. This very popular model saw nearly 700 units produced up to 1965.

This series II Type 362 is mounted on a self-propelled carrier and can be seen working in an oil refinery in 1959.

Another Type 362 being used in a unique fashion. This 1952 shot shows the machine set up as a slackline crane handling a concrete bucket for Johnson Bros. Company Ltd., Brantford, Ontario, Canada. Note the extra stability provided from the guy line.

The Marion 2 1/2-cubic-yard 93-M was built from 1946 through 1975, with 364 units built. This machine is placing sand drains in Portland, Maine. The mandrel is driven into the ground and then sand is poured down through the mandrel. After applying steam pressure on the sand, the mandrel is withdrawn, leaving the sand in place.

Boasting a 50-ton lifting capacity, the 93-M is unloading a full bunk of pulpwood in one pick!

99

A beautiful photograph of a 93-M on the erecting floor of the Marion Power Shovel Company. Modern features include hook rollers and a full-revolving dragline fairlead.

Also introduced in 1946, this 4 1/2-cubic-yard 111-M dragline is loading a fleet of Caterpillar DW20 scrapers.

This 111-M shovel is one of 379 units built by 1974, when production finally ended. Equally popular as a shovel or dragline, the 111-M was a favorite in coal stripping operations.

The 1-cubic-yard 43-M was first offered in 1949. One of the most popular of Marion's modern fleet, 437 were built by 1965.

The 43-M backhoe pictured here has the standard gooseneck hoe boom. A longer boom option gave this machine a respectable 25-foot digging depth.

When equipped with 65 feet of boom as shown in this September 1959 photograph, the 43-M dragline could dig to a depth of 48 feet.

The 43-M truck crane was first offered with a 27-ton capacity, but that was increased over subsequent models, culminating in a 40-ton capacity on a four-axle carrier.

The 3-cubic-yard 101-M was introduced in 1953 and was Marion's first machine to be offered with a torque converter as standard equipment. One hundred twelve machines were built through 1975.

The 2-cubic-yard 83-M came on the scene in 1954 and was offered until 1957. The twin dipper sticks, large diameter hoist sheaves, and box construction boom are clearly evident.

In 1956 Marion offered its new 3/4-cubic-yard 35-M. The 35-M backhoe could dig 19-feet deep and could handle loads up to 5 tons.

This Marion 3/4-cubic-yard 35-M shovel is working for the Alabama State Highway Department in August 1960. This popular-sized machine featured a one-piece machinery deck and welded sideframes.

The Marion 37-M truck crane came on the scene in 1960 as a 35-ton machine. This photograph shows its ability to fold and carry 100 feet of boom and jib. The 37-M also featured pin-connected boom sections, retractable gantry, and removable counterweights.

The Marion 37-M truck crane was also Marion's only mobile tower crane. This photograph shows an experimental traveling outrigger system and closed circuit TV monitor.

The year 1960 was a big year for new Marion machines. Seven "M" series machines were introduced, including the 35-ton, 1 1/4-cubic-yard 45-M. The 43-M, 45-M, and 47-M all shared the same platform. The power load-lowering option made handling a concrete bucket "user friendly"!

The Marion 47-M was offered only as a lift crane during its 4-year tenure. Rated at 45 tons, this workhorse could also handle up to 160 feet of boom and jib combination.

The 65-M, 75-M, and 77-M were also introduced in 1960. These machines shared a similar platform of deck machinery. This composite view shows the 1 1/2-cubic-yard 65-M with its various front ends. Only 29 machines were built in this series.

Sometime around 1886, the Marion Steam Shovel Company entered into the production of railroad "ditchers." These shovels were used to dig and maintain the ditches that paralleled the track and were so crucial to drainage. The boom swung on a circle similar to the railroad shovel, but the upperworks could be rotated to about a 45-degree angle, as seen in this early photograph. This 1/2-cubic-yard machine is owned by the Texas Central Railway.

By 1909, the 1/2-cubic-yard Model 15 Ditcher was introduced by Marion, replacing the original ditcher model. This machine was fully revolving and could dig three to four feet below the rails. Pictured above is a machine owned by the Canadian Pacific Railway, one of 20 Marion had built by 1919.

The Marion Model 17 Ditcher was only offered for a little over one year, from 1910 to 1911, and only five examples were built. This factory photograph shows the "racer" undercarriage, the means by which the machine traversed the rails on top of the string of flatcars, thus enabling it to load each car in succession.

The 5/8-cubic-yard Model 18 Ditcher was offered from 1912 through 1919, with nine units being constructed. This machine is swinging the track sections it uses for traveling the length of the flatcars.

The 3/4-cubic-yard Model 21 Ditcher was the last produced by Marion and was nothing more than the standard Model 21 steam shovel equipped with the dual-gauge undercarriage.

Marion introduced the hugely successful Barnhart Log Loader in 1888, and had sold 160 of these machines by 1906, when the Model 10 Log Loader was brought on the scene. Used to load logs onto small flatcars or skeleton cars, the loader sat on a sled and winched its way over the flatcars as they were loaded. These were Marion's first full-revolving machines. The boom was at a fixed angle, but the machine had a "live" circle and endless chain for rotating. (The skidding sled and rotating chain are visible in this photograph.) Sixty-nine Model 10 units were built by 1927.

The Model 12 Log Loader was smaller than the Model 10 and was introduced the same year. Its "claim to fame" was its pinion and swing gear arrangement used for rotating the upperworks. This photograph shows a 1909 Model 12 sitting on erecting trucks. Only 20 of these machines were built by 1916.

A post-WWII Marion Type 362 crawler equipped with a heel boom. The tongs are attached near the end of the log, and the log is hoisted until it butts, or heels, against the underside of the boom, preventing it from swinging out of control during loading.

In the 1950s, Marion began a vigorous campaign to supply the logging industry with modern log loaders. This 43-M truck-mounted logger features a heel boom front end, standard on most loggers, and hydraulic outriggers.

This 1961 photograph features a truck-mounted 47-M logger with an elevated cab. Other desirable features on these machines included power load lowering on the main hoist and a third drum.

Index by Model

Barnhart RR	p. 6
Barnhart Special RR	p. 9
A (RR)	pp. 7, 8
AA (RR)	pp. 11, 12
C (RR)	p. 10
G (RR)	pp. 14, 15
H (RR)	p. 13
K (RR)	p. 16
Barnhart Ditcher RR	p. 116
7	pp. 74, 75, 76, 77, 78
10	p. 121
12	p. 122
15	p. 117
17	p. 118
18	p. 119
20 RR	pp. 18, 19
21	pp. 59, 60, 61, 62, 63, 120
25 RR	pp. 31, 32
28	pp. 42, 43, 44, 45
30	pp. 38, 39
31	pp. 47, 48
32	pp. 64, 65, 66
35	pp. 40, 41
35-M	pp. 109, 110
36	pp. 50, 51, 52, 53
37	pp. 67, 68, 69, 70
37-M	pp. 111, 112
39-A	p. 93
40 RR	pp. 25, 26
41 RR	p. 34
43-M	pp. 103, 104, 105, 106, 124
45-M	p. 113
47-M	pp. 114, 125
50 RR	p. 23
51 RR	p. 37
60 RR	p. 20
61 RR	p. 35
65-M	p. 115
70 RR	p. 36
75 RR	p. 24
76 RR	p. 33
80 RR	p. 17
83-M	p. 108
91 RR	pp. 21, 22
92 RR	pp. 27, 28, 29
93-M	pp. 98, 99, 100
100 RR	p. 30
101-M	p. 107
111-M	pp. 101, 102
120	pp. 90, 91
220	p. 54
221	p. 58
222	pp. 72, 73
250	p. 46
251	p. 55
260	p. 49
261	p. 56
262	p. 71
281	p. 57
331	p. 92
340	p. 89
362	pp. 95, 96, 97, 123
372	p. 94
440	p. 79
450	pp. 81, 82, 83, 84
460	p. 80
480	pp. 85, 86, 87, 88

MORE TITLES FROM ICONOGRAFIX:

AMERICAN CULTURE
AMERICAN SERVICE STATIONS 1935-1943 PHOTO ARCHIVE	ISBN 1-882256-27-1
COCA-COLA: A HISTORY IN PHOTOGRAPHS 1930-1969	ISBN 1-882256-46-8
COCA-COLA: ITS VEHICLES IN PHOTOGRAPHS 1930-1969	ISBN 1-882256-47-6
PHILLIPS 66 1945-1954 PHOTO ARCHIVE	ISBN 1-882256-42-5
RVs & CAMPERS 1900-2000: AN ILLUSTRATED HISTORY	ISBN 1-58388-064-X

AUTOMOTIVE
AMX PHOTO ARCHIVE: FROM CONCEPT TO REALITY	ISBN 1-58388-062-3
CADILLAC 1948-1964 PHOTO ALBUM	ISBN 1-882256-83-2
CAMARO 1967-2000 PHOTO ARCHIVE	ISBN 1-58388-032-1
CHEVROLET STATION WAGONS 1946-1966 PHOTO ARCHIVE	ISBN 1-58388-069-0
CLASSIC AMERICAN LIMOUSINES 1955-2000 PHOTO ARCHIVE	ISBN 1-58388-041-0
CORVAIR by CHEVROLET EXP. & PROD. CARS 1957-1969 LUDVIGSEN LIBRARY SERIES	ISBN 1-58388-058-5
CORVETTE THE EXOTIC EXPERIMENTAL CARS, LUDVIGSEN LIBRARY SERIES	ISBN 1-58388-017-8
CORVETTE PROTOTYPES & SHOW CARS PHOTO ALBUM	ISBN 1-882256-77-8
EARLY FORD V-8S 1932-1942 PHOTO ALBUM	ISBN 1-882256-97-2
IMPERIAL 1955-1963 PHOTO ARCHIVE	ISBN 1-882256-22-0
IMPERIAL 1964-1968 PHOTO ARCHIVE	ISBN 1-882256-23-9
LINCOLN MOTOR CARS 1920-1942 PHOTO ARCHIVE	ISBN 1-882256-57-3
LINCOLN MOTOR CARS 1946-1960 PHOTO ARCHIVE	ISBN 1-882256-58-1
PACKARD MOTOR CARS 1935-1942 PHOTO ARCHIVE	ISBN 1-882256-44-1
PACKARD MOTOR CARS 1946-1958 PHOTO ARCHIVE	ISBN 1-882256-45-X
PONTIAC DREAM CARS, SHOW CARS & PROTOTYPES 1928-1998 PHOTO ALBUM	ISBN 1-882256-93-X
PONTIAC FIREBIRD TRANS-AM 1969-1999 PHOTO ALBUM	ISBN 1-882256-95-6
PONTIAC FIREBIRD 1967-2000 PHOTO HISTORY	ISBN 1-58388-028-3
STRETCH LIMOUSINES 1928-2001 PHOTO ARCHIVE	ISBN 1-58388-070-4
STUDEBAKER 1933-1942 PHOTO ARCHIVE	ISBN 1-882256-24-7
ULTIMATE CORVETTE TRIVIA CHALLENGE	ISBN 1-58388-035-6

BUSES
BUSES OF MOTOR COACH INDUSTRIES 1932-2000 PHOTO ARCHIVE	ISBN 1-58388-039-9
FLXIBLE TRANSIT BUSES 1953-1995 PHOTO ARCHIVE	ISBN 1-58388-053-4
GREYHOUND BUSES 1914-2000 PHOTO ARCHIVE	ISBN 1-58388-027-5
MACK® BUSES 1900-1960 PHOTO ARCHIVE*	ISBN 1-58388-020-8
TRAILWAYS BUSES 1936-2001 PHOTO ARCHIVE	ISBN 1-58388-029-1
TROLLEY BUSES 1913-2001 PHOTO ARCHIVE	ISBN 1-58388-057-7
YELLOW COACH BUSES 1923-1943 PHOTO ARCHIVE	ISBN 1-58388-054-2

EMERGENCY VEHICLES
AMERICAN LAFRANCE 700 SERIES 1945-1952 PHOTO ARCHIVE	ISBN 1-882256-90-5
AMERICAN LAFRANCE 700 SERIES 1945-1952 PHOTO ARCHIVE VOLUME 2	ISBN 1-58388-025-9
AMERICAN LAFRANCE 700 & 800 SERIES 1953-1958 PHOTO ARCHIVE	ISBN 1-882256-91-3
AMERICAN LAFRANCE 900 SERIES 1958-1964 PHOTO ARCHIVE	ISBN 1-58388-002-X
CROWN FIRECOACH 1951-1985 PHOTO ARCHIVE	ISBN 1-58388-047-X
CLASSIC AMERICAN AMBULANCES 1900-1979 PHOTO ARCHIVE	ISBN 1-882256-94-8
CLASSIC AMERICAN FUNERAL VEHICLES 1900-1980 PHOTO ARCHIVE	ISBN 1-58388-016-X
CLASSIC SEAGRAVE 1935-1951 PHOTO ARCHIVE	ISBN 1-58388-034-8
FIRE CHIEF CARS 1900-1997 PHOTO ALBUM	ISBN 1-882256-87-5
HEAVY RESCUE TRUCKS 1931-2000 PHOTO GALLERY	ISBN 1-58388-045-3
INDUSTRIAL AND PRIVATE FIRE APPARATUS 1925-2001 PHOTO ARCHIVE	ISBN 1-58388-049-6
LOS ANGELES CITY FIRE APPARATUS 1953 - 1999 PHOTO ARCHIVE	ISBN 1-58388-012-7
MACK MODEL C FIRE TRUCKS 1957-1967 PHOTO ARCHIVE*	ISBN 1-58388-014-3
MACK MODEL L FIRE TRUCKS 1940-1954 PHOTO ARCHIVE*	ISBN 1-882256-86-7
MAXIM FIRE APPARATUS 1914-1989 PHOTO ARCHIVE	ISBN 1-58388-050-X
NAVY & MARINE CORPS FIRE APPARATUS 1836-2000 PHOTO GALLERY	ISBN 1-58388-031-3
POLICE CARS: RESTORING, COLLECTING & SHOWING AMERICA'S FINEST SEDANS	ISBN 1-58388-046-1
SEAGRAVE 70TH ANNIVERSARY SERIES PHOTO ARCHIVE	ISBN 1-58388-001-1
TASC FIRE APPARATUS 1946-1985 PHOTO ARCHIVE	ISBN 1-58388-065-8
VOLUNTEER & RURAL FIRE APPARATUS PHOTO GALLERY	ISBN 1-58388-005-4
W.S. DARLEY & CO. FIRE APPARATUS 1908-2000 PHOTO ACHIVE	ISBN 1-58388-061-5
WARD LAFRANCE FIRE TRUCKS 1918-1978 PHOTO ARCHIVE	ISBN 1-58388-013-5
WILDLAND FIRE APPARATUS 1940-2001 PHOTO GALLERY	ISBN 1-58388-056-9
YOUNG FIRE EQUIPMENT 1932-1991 PHOTO ARCHIVE	ISBN 1-58388-015-1

RACING
CHAPARRAL CAN-AM RACING CARS FROM TEXAS LUDVIGSEN LIBRARY SERIES	ISBN 1-58388-066-6
DRAG RACING FUNNY CARS OF THE 1970s PHOTO ARCHIVE	ISBN 1-58388-068-2
EL MIRAGE IMPRESSIONS: DRY LAKES LAND SPEED RACING	ISBN 1-58388-059-3
GT40 PHOTO ARCHIVE	ISBN 1-882256-64-6
INDY CARS OF THE 1950s, LUDVIGSEN LIBRARY SERIES	ISBN 1-58388-018-6
INDY CARS OF THE 1960s, LUDVIGSEN LIBRARY SERIES	ISBN 1-58388-052-6
INDIANAPOLIS RACING CARS OF FRANK KURTIS 1941-1963 PHOTO ARCHIVE	ISBN 1-58388-026-7
JUAN MANUEL FANGIO WORLD CHAMPION DRIVER SERIES PHOTO ALBUM	ISBN 1-882256-08-9
LE MANS 1950 PHOTO ARCHIVE THE BRIGGS CUNNINGHAM CAMPAIGN	ISBN 1-882256-21-2
MARIO ANDRETTI WORLD CHAMPION DRIVER SERIES PHOTO ALBUM	ISBN 1-58388-009-7
MERCEDES-BENZ 300SL RACING CARS 1952-1953 LUDVIGSEN LIBRARY SERIES	ISBN 1-58388-067-4
NOVI V-8 INDY CARS 1941-1965 LUDVIGSEN LIBRARY SERIES	ISBN 1-58388-037-2
SEBRING 12-HOUR RACE 1970 PHOTO ARCHIVE	ISBN 1-882256-20-4
VANDERBILT CUP RACE 1936 & 1937 PHOTO ARCHIVE	ISBN 1-882256-66-2

RAILWAYS
CHICAGO, ST. PAUL, MINNEAPOLIS & OMAHA RAILWAY 1880-1940 PHOTO ARCHIVE	ISBN 1-882256-67-0
CHICAGO & NORTH WESTERN RAILWAY 1975-1995 PHOTO ARCHIVE	ISBN 1-882256-76-X
GREAT NORTHERN RAILWAY 1945-1970 PHOTO ARCHIVE	ISBN 1-882256-56-5
GREAT NORTHERN RAILWAY 1945-1970 VOL 2 PHOTO ARCHIVE	ISBN 1-882256-79-4
ILLINOIS CENTRAL RAILROAD 1854-1960 PHOTO ARCHIVE	ISBN 1-58388-063-1
MILWAUKEE ROAD 1850-1960 PHOTO ARCHIVE	ISBN 1-882256-61-1
MILWAUKEE ROAD DEPOTS 1856-1954 PHOTO ARCHIVE	ISBN 1-58388-040-2
SHOW TRAINS OF THE 20TH CENTURY	ISBN 1-58388-030-5
SOO LINE 1975-1992 PHOTO ARCHIVE	ISBN 1-882256-68-9
TRAINS OF THE TWIN PORTS, DULUTH-SUPERIOR IN THE 1950s PHOTO ARCHIVE	ISBN 1-58388-003-8
TRAINS OF THE CIRCUS 1872-1956	ISBN 1-58388-024-0
TRAINS of the UPPER MIDWEST PHOTO ARCHIVE STEAM&DIESEL in the1950S&1960S	ISBN 1-58388-036-4
WISCONSIN CENTRAL LIMITED 1987-1996 PHOTO ARCHIVE	ISBN 1-882256-75-1
WISCONSIN CENTRAL RAILWAY 1871-1909 PHOTO ARCHIVE	ISBN 1-882256-78-6

TRUCKS
BEVERAGE TRUCKS 1910-1975 PHOTO ARCHIVE	ISBN 1-882256-60-3
BROCKWAY TRUCKS 1948-1961 PHOTO ARCHIVE*	ISBN 1-882256-55-7
CHEVROLET EL CAMINO PHOTO HISTORY INCL GMC SPRINT & CABALLERO	ISBN 1-58388-044-5
CIRCUS AND CARNIVAL TRUCKS 1923-2000 PHOTO ARCHIVE	ISBN 1-58388-048-8
DODGE PICKUPS 1939-1978 PHOTO ALBUM	ISBN 1-882256-82-4
DODGE POWER WAGONS 1940-1980 PHOTO ARCHIVE	ISBN 1-882256-89-1
DODGE POWER WAGON PHOTO HISTORY	ISBN 1-58388-019-4
DODGE RAM TRUCKS 1994-2001 PHOTO HISTORY	ISBN 1-58388-051-8
DODGE TRUCKS 1929-1947 PHOTO ARCHIVE	ISBN 1-882256-36-0
DODGE TRUCKS 1948-1960 PHOTO ARCHIVE	ISBN 1-882256-37-9
FORD HEAVY-DUTY TRUCKS 1948-1998 PHOTO HISTORY	ISBN 1-58388-043-7
JEEP 1941-2000 PHOTO ARCHIVE	ISBN 1-58388-021-6
JEEP PROTOTYPES & CONCEPT VEHICLES PHOTO ARCHIVE	ISBN 1-58388-033-X
LOGGING TRUCKS 1915-1970 PHOTO ARCHIVE	ISBN 1-882256-59-X
MACK MODEL AB PHOTO ARCHIVE*	ISBN 1-882256-18-2
MACK AP SUPER-DUTY TRUCKS 1926-1938 PHOTO ARCHIVE*	ISBN 1-882256-54-9
MACK MODEL B 1953-1966 VOL 1 PHOTO ARCHIVE*	ISBN 1-882256-19-0
MACK MODEL B 1953-1966 VOL 2 PHOTO ARCHIVE*	ISBN 1-882256-34-4
MACK EB-EC-ED-EE-EF-EG-DE 1936-1951 PHOTO ARCHIVE*	ISBN 1-882256-29-8
MACK EH-EJ-EM-EQ-ER-ES 1936-1950 PHOTO ARCHIVE*	ISBN 1-882256-39-5
MACK FC-FCSW-NW 1936-1947 PHOTO ARCHIVE*	ISBN 1-882256-28-X
MACK FG-FH-FJ-FK-FN-FP-FT-FW 1937-1950 PHOTO ARCHIVE*	ISBN 1-882256-35-2
MACK LF-LH-LJ-LM-LT 1940-1956 PHOTO ARCHIVE*	ISBN 1-882256-38-7
MACK TRUCKS PHOTO GALLERY*	ISBN 1-882256-88-3
NEW CAR CARRIERS 1910-1998 PHOTO ALBUM	ISBN 1-882256-98-0
PLYMOUTH COMMERCIAL VEHICLES PHOTO ARCHIVE	ISBN 1-58388-004-6
REFUSE TRUCKS PHOTO ARCHIVE	ISBN 1-58388-042-9
STUDEBAKER TRUCKS 1927-1940 PHOTO ARCHIVE	ISBN 1-882256-40-9
STUDEBAKER TRUCKS 1941-1964 PHOTO ARCHIVE	ISBN 1-882256-41-7
WHITE TRUCKS 1900-1937 PHOTO ARCHIVE	ISBN 1-882256-80-8

TRACTORS & CONSTRUCTION EQUIPMENT
CASE TRACTORS 1912-1959 PHOTO ARCHIVE	ISBN 1-882256-32-8
CATERPILLAR PHOTO GALLERY	ISBN 1-882256-70-0
CATERPILLAR POCKET GUIDE THE TRACK-TYPE TRACTORS 1925-1957	ISBN 1-58388-022-4
CATERPILLAR D-2 & R-2 PHOTO ARCHIVE	ISBN 1-882256-99-9
CATERPILLAR D-8 1933-1974 PHOTO ARCHIVE INCLUDING DIESEL 75 & RD-8	ISBN 1-882256-96-4
CATERPILLAR MILITARY TRACTORS VOLUME 1 PHOTO ARCHIVE	ISBN 1-882256-16-6
CATERPILLAR MILITARY TRACTORS VOLUME 2 PHOTO ARCHIVE	ISBN 1-882256-17-4
CATERPILLAR SIXTY PHOTO ARCHIVE	ISBN 1-882256-05-0
CATERPILLAR TEN PHOTO ARCHIVE INCLUDING 7C FIFTEEN & HIGH FIFTEEN	ISBN 1-58388-011-9
CATERPILLAR THIRTY PHOTO ARCHIVE 2ND ED. INC. BEST THIRTY, 6G THIRTY & R-4	ISBN 1-58388-006-2
CLETRAC AND OLIVER CRAWLERS PHOTO ARCHIVE	ISBN 1-58388-043-3
CLASSIC AMERICAN STEAMROLLERS 1871-1935 PHOTO ARCHIVE	ISBN 1-58388-038-0
FARMALL CUB PHOTO ARCHIVE	ISBN 1-882256-71-9
FARMALL F– SERIES PHOTO ARCHIVE	ISBN 1-882256-02-6
FARMALL MODEL H PHOTO ARCHIVE	ISBN 1-882256-03-4
FARMALL MODEL M PHOTO ARCHIVE	ISBN 1-882256-15-8
FARMALL REGULAR PHOTO ARCHIVE	ISBN 1-882256-14-X
FARMALL SUPER SERIES PHOTO ARCHIVE	ISBN 1-882256-49-2
FORDSON 1917-1928 PHOTO ARCHIVE	ISBN 1-882256-33-6
HART-PARR PHOTO ARCHIVE	ISBN 1-882256-08-5
HOLT TRACTORS PHOTO ARCHIVE	ISBN 1-882256-10-7
INTERNATIONAL TRACTRACTOR PHOTO ARCHIVE	ISBN 1-58388-048-4
INTERNATIONAL TD CRAWLERS 1933-1962 PHOTO ARCHIVE	ISBN 1-882256-72-7
JOHN DEERE MODEL A PHOTO ARCHIVE	ISBN 1-882256-12-3
JOHN DEERE MODEL B PHOTO ARCHIVE	ISBN 1-882256-01-8
JOHN DEERE MODEL D PHOTO ARCHIVE	ISBN 1-882256-00-X
JOHN DEERE 30 SERIES PHOTO ARCHIVE	ISBN 1-882256-13-1
MARION CONSTRUCTION MACHINERY 1884 - 1975 PHOTO ARCHIVE	ISBN 1-58388-060-7
MINNEAPOLIS-MOLINE U-SERIES PHOTO ARCHIVE	ISBN 1-882256-07-7
OLIVER TRACTORS PHOTO ARCHIVE	ISBN 1-882256-09-3
RUSSELL GRADERS PHOTO ARCHIVE	ISBN 1-882256-11-5
TWIN CITY TRACTOR PHOTO ARCHIVE	ISBN 1-882256-06-9

*This product is sold under license from Mack Trucks, Inc. Mack is a registered Trademark of Mack Trucks, Inc. All rights reserved.

All Iconografix books are available from direct mail specialty book dealers and bookstores worldwide, or can be ordered from the publisher. For book trade and distribution information or to add your name to our mailing list and receive a **FREE CATALOG** contact:

Iconografix, PO Box 446, Hudson, Wisconsin, 54016   Telephone: (715) 381-9755, (800) 289-3504 (USA), Fax: (715) 381-9756

MORE GREAT BOOKS FROM ICONOGRAFIX

CATERPILLAR D-2 & R-2 PHOTO ARCHIVE
ISBN 1-882256-99-9

CATERPILLAR D-8 1933-1974 PHOTO ARCHIVE INCLUDING DIESEL 75 & RD-8
ISBN 1-882256-96-4

CATERPILLAR THIRTY PHOTO ARCHIVE 2ND ED. INC. BEST THIRTY, 6G THIRTY & R-4
ISBN 1-58388-006-2

HOLT TRACTORS PHOTO ARCHIVE
ISBN 1-882256-10-7

CATERPILLAR POCKET GUIDE THE TRACK-TYPE TRACTORS 1925-1957
ISBN 1-58388-022-4

CLASSIC AMERICAN STEAMROLLERS 1871-1935 PHOTO ARCHIVE
ISBN 1-58388-038-0

CATERPILLAR PHOTO GALLERY
ISBN 1-882256-70-0

**ICONOGRAFIX, INC.
P.O. BOX 446, DEPT BK,
HUDSON, WI 54016
FOR A FREE CATALOG CALL:
1-800-289-3504**

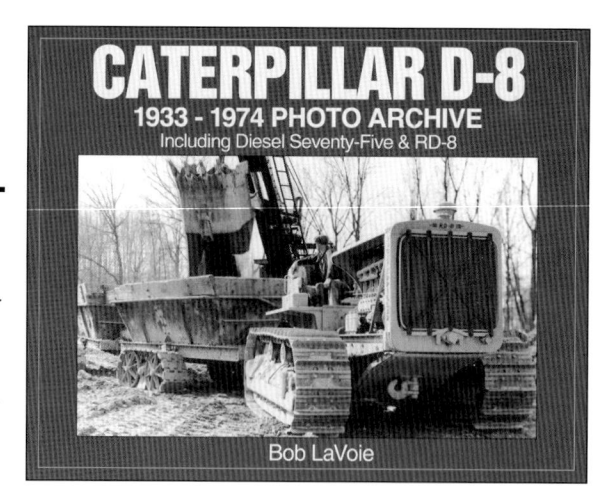

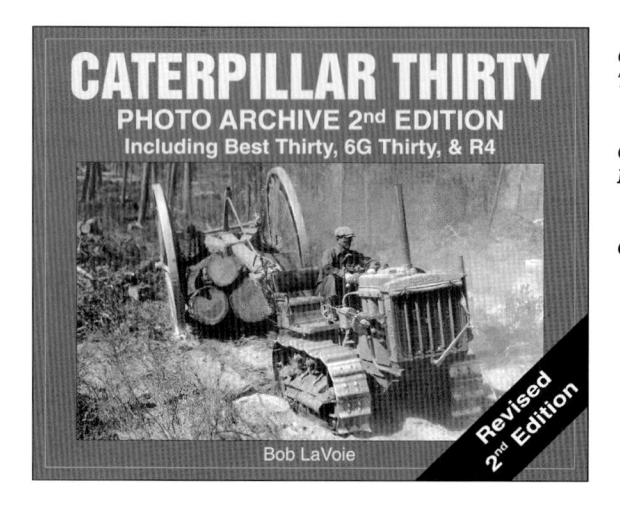

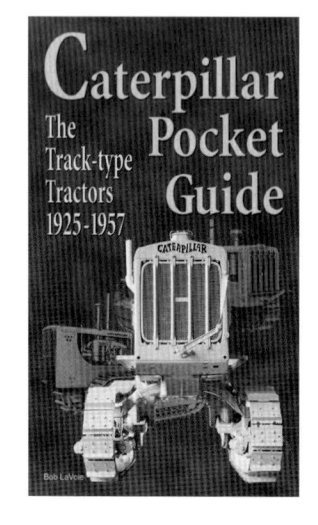

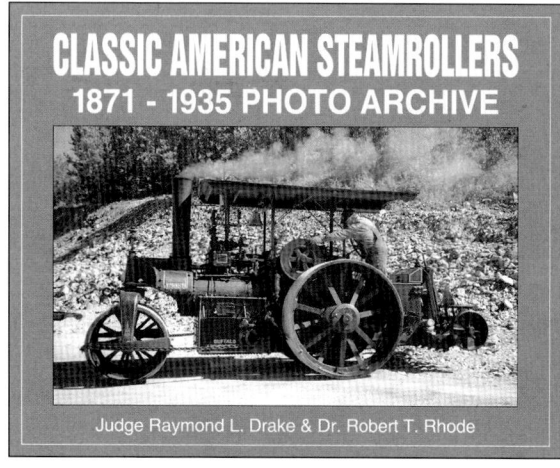